Ricardo Boscaini
José Roberto Chaves Neto
Ivan Dressler da Costa

Physiological and sanitary quality of soybean seeds

Ricardo Boscaini
José Roberto Chaves Neto
Ivan Dressler da Costa

Physiological and sanitary quality of soybean seeds

Physiology and Health

ScienciaScripts

Imprint

Any brand names and product names mentioned in this book are subject to trademark, brand or patent protection and are trademarks or registered trademarks of their respective holders. The use of brand names, product names, common names, trade names, product descriptions etc. even without a particular marking in this work is in no way to be construed to mean that such names may be regarded as unrestricted in respect of trademark and brand protection legislation and could thus be used by anyone.

Cover image: www.ingimage.com

This book is a translation from the original published under ISBN 978-613-9-62638-0.

Publisher:
Sciencia Scripts
is a trademark of
Dodo Books Indian Ocean Ltd. and OmniScriptum S.R.L publishing group

120 High Road, East Finchley, London, N2 9ED, United Kingdom
Str. Armeneasca 28/1, office 1, Chisinau MD-2012, Republic of Moldova, Europe
Printed at: see last page
ISBN: 978-620-7-74774-0

SUMMARY

SUMMARY

PHYSIOLOGICAL AND SANITARY QUALITY OF SOYBEAN SEEDS

Among the factors that help the performance of the crop in the field and consequently the achievement of a uniform crop, with an ideal plant population, which often depends on certain practices, we can highlight high quality seeds together with the treatment of products for them, enabling them to perform better in the field. The aim of this study was to assess the physiological quality and health of soya bean seeds from plants subjected to foliar applications of different fungicides. This study was conducted at the Federal University of Santa Maria, in the Central Depression climate region of the state of Rio Grande do Sul, at an altitude of 116 meters (m), latitude 29°42'24" S and longitude 53°48'42" W. The cultivar used was BMX Apolo RR, where each plot consisted of $3m^2$ of area for evaluation. Four applications were made, at V5, R1, R1 + 14 days and R1 + 28 days. The experimental design was entirely randomized, with a factorial scheme of 8 x 4 (8 treatments x 4 replications). For the physiological and health assessment, the seeds were subjected to the following tests: germination, seedling length, pathology and thousand-grain mass. We can conclude that the chemical treatment with picoxystrobin + cyproconazole was the most efficient for the physiological evaluation of soybean seeds. The chemical treatment metconazole + pyraclostrobin provided greater control of seed-borne pathogens.

Keywords: physiology; *Glycine max*; health; vigor test.

1. INTRODUCTION

Currently, soya is the main product of Brazilian agriculture, placing Brazil in second place in world production, occupying a prominent place in the agribusiness scenario as a major producer and exporter of this crop. This sector accounts for approximately 37% of formal jobs and around 25% of the national GDP. Despite the productive potential of Brazilian cultivars, some abiotic and biotic factors can interfere with a reduction in crop productivity.

Seed quality involves genetic, physical, physiological and health aspects. The sanitary condition is extremely important for the production of high quality plants, because since seeds are effective in spreading phytopathogenic agents, these can be taken to the field, causing a reduction in germination and vigor, affecting the crop's stand and quantitatively and qualitatively reducing production.

Among the factors that help the performance of the crop in the field and consequently the achievement of a uniform crop, with an ideal plant stand, which often depends on some practices, we can highlight high quality seeds together with the treatment of products for them, enabling them to perform better in the field (MERTZ et al., 2009).

Due to the great importance of soybeans to Brazilian agribusiness, research is needed into the factors that affect the maximum expression of their productive potential. One of the aspects that affect the performance of most crops is the occurrence of diseases, which play an important role in reducing the quality and productivity of food crops, thus increasing their production costs (BARROS et al., 2005).

An effective control measure that prevents the entry and establishment of various pathogens in soybean seeds is chemical treatment, aimed not only at preserving seed quality, but also at improving germination, as well as protecting seedlings from pathogens residing in the soil and under adverse conditions (DHINGRA, 2005).

Some studies show that the effect of fungicides used in seed treatment on physiological quality can vary depending on the product used, thus causing increases (PEREIRA et al., 2007) or decreases (GIANASI et al., 2000) in seed vigor and germination.

In the soybean crop, there are several pathogens that cause damage, among which *Colletotrichum truncatum, Cercospora kikuchii, Phomopsis sp., Fusarium sp., Aspergillus sp. and Penicillium sp. stand* out, which cause serious damage to the physiological quality of the seeds, generating significant losses in the establishment of the initial plant stand, in other words, causing a reduction in germination (KROHN; MALAVASI, 2004).

In view of the above, the aim of this study was to evaluate the physiological quality and health of soybean seeds from plants subjected to foliar applications of different fungicides.

2. LITERATURE REVIEW

The aim of this literature review is to provide a theoretical basis for the topic proposed in this work. In this way, this chapter presents a literature review on the soybean crop (*Glycine max*), through a brief history of the general aspects of the crop, production technology, seed pathology, physiological quality, economic damage and chemical groups. Finally, we present existing research on the subject.

2.1 SOYBEAN CULTIVATION IN BRAZIL

Soybean (Glicine max l.) is an annual legume (90 to 160 days) that originated in the Far East, in China, and has been cultivated for thousands of years. In the 1920s, American farmers began growing soybeans on a large scale, which were mainly used as an input for animal feed (HIN, 2002).

In Brazil, soybeans were introduced by Gustavo Dutra in Bahia in 1882, without success, as the germplasm brought from the United States did not adapt to the low latitude conditions of the region (12°S). Gustavo Dutra, then a professor at the Bahia School of Agronomy, began to carry out the first studies evaluating the cultivars introduced from that country. Cultivated for the first time in São Paulo (latitude 23°S), by Daffert, in 1892, at the Agronomic Institute of Campinas. In São Paulo, new materials were tested and in this case it was relatively successful in producing hay and grain (Sediyama et al., 1985; Kiihl, 2006; Seixas et al., 2006).

The first record of soybean cultivation in Brazil was in the municipality of Santa Rosa, RS, in 1914. But it gained economic importance from the 1940s onwards, with its first national statistical record in 1941, in the Anuàrio Agricola do RS. In 1949, with a production of 25,000 tons, Brazil was listed for the first time as a soybean producer in international statistics (EMBRAPA SOJA, 2003).

In 1900, it was introduced to Rio Grande do Sul. In this state, detailed descriptions of its botanical and cultural aspects were carried out in 1901, as well as experimental evaluations in 1921, while its cultivation for agricultural purposes began in 1924. In 1931, the cultivated variety Amarela-rio grande was introduced which, for many years, accounted for more than 95% of soybean production in Rio Grande do Sul, and in 1965, out of a total of 350,000 hectares under cultivation, 70% of the area was cultivated with this variety (FERREIRA, 2002).

The crop has seen a growth in production and an increase in competitive capacity within the Brazilian market over the years. This evolution has always been associated with technological and scientific advances in soybean production (EMBRAPA, 2004). Over the last 30 years, soybeans have been the fastest-growing crop in Brazil, accounting for 49% of the total area under crops in Brazil (MAPA, 2012).

Soybeans (*Glycine max* L.) are the most economically significant crop in Brazil, with a production of around 96.203 million tons in an area of approximately 31.940 million hectares, making it the world's second largest producer (CONAB, 2015).

Soybeans in Brazil are predominantly used for processing into oil and protein. The processed protein (cake or bran) is used as a protein supplement in animal feed. This bran is roasted/heated to the point of inactivating the anti-nutritional factors naturally present in soy (trypsin inhibitor, stachyose, raffinose, phytate). The United States and Brazil are the largest suppliers of soybean meal on the world market. Soybean oil is widely used by the Brazilian population and contains only traces of protein. This oil is consumed directly or in processed foods such as margarine (SOJA, 2013).

Scientific research has been collaborating with the satisfactory results of Brazilian productivity gains, with studies focused on genetics and plant breeding, soils and nutrition, phytopathology, agricultural machinery, among others, so that they can all, in some integrated or multidisciplinary way, achieve higher levels of productivity (NETO et al., 2015).

According to Southgate (2009), with the increase in population since the middle of the 20th century, the demand for food has increased. In Brazil, this has led to major changes in agricultural production. The policy of stimulating rural credit, together with new technologies, has boosted various crops, especially those destined for export, in this case soya being one of these crops, with the aim of increasing production to meet existing demand.

Many diseases caused by fungi, bacteria, nematodes and viruses have already been identified in Brazil. Usually fungal diseases are the most recurrent and, consequently, the ones that cause the most damage to crops. With the expansion of soya into new areas and as a result of monoculture, the number of pathogens continues to increase. Annual production losses due to diseases are estimated at between 15% and 20%, but some diseases can cause losses of up to 100% (EMBRAPA SOJA, 2003).

Diseases are identified as one of the main factors limiting high yields of soybeans, affecting crop productivity and causing damage to seed quality. Diseases of various etiologies have already been identified in Brazil. Phytosanitary problems have increased with the expansion of soybeans into new areas, as a result of monoculture and the introduction of new pathogens. Although annual production losses due to diseases are estimated at around 15% to 20%, some diseases can cause losses of almost 100% (ALMEIDA et al., 2005).

The fungal diseases that affect soybeans deserve special attention, not only because of their greater number, but also because of the damage they cause to the yield and quality of the seeds. Numerous microorganisms have been identified in soybean seeds, but few are considered economically important (HENNING, 2004).

2.2 SEED PRODUCTION TECHNOLOGY

The seed production process has several phases, including research, improvement, production, certification, post-harvest maintenance and, if the seeds are to be sold, marketing (FAO, 2011). Obtaining high-quality seeds is the priority goal of the seed production process. In this context, processing is an essential stage in seed production, as the seed lot needs to be properly processed and handled, otherwise the previous efforts in the seed production phase could be nullified (FERREIRA, 2010).

We can say that the soybean seed production process is similar to the grain production process in terms of the way it is carried out, but some details differ between them: choosing an area with a higher altitude (above 700 meters), in tropical regions, due to the milder temperatures, and finally the field inspection to determine the need for *roguing*.

Federal Law 10.711 of August 5, 2003, in its first article, known as the Seed Law, aims to clarify the quality of the plant multiplication and reproduction material produced, as well as its identity, which is marketed throughout the country (BRASIL, 2003).

According to statistics from the Brazilian Seed and Seedling Association (ABRASEM), Brazil uses 64% of the seed utilization rate (SUT), which means a demand for approximately 814,000 tons of soybean seeds. However, our consumption potential is around 1,270,000 tons of soybean seeds. Mato Grosso has the highest rate of soybean seed use compared to the other states in the union at 83%, while the lowest rate is Rio Grande do Sul with 40% (ABRASEM, 2011).

In order to produce seed legally, the producer and the Ministry of Agriculture, Livestock and Food Supply (MAPA) must have certain controls in place throughout the entire production period, and each approved batch must receive documentation in accordance with the seed law: 1- A seed analysis bulletin certified by the MAPA and entered in the National Seed Register

(RENASEM), for all seed categories; 2- Certification established by the MAPA and signed by a Technical Officer, certifying the Basic, Certifica 1 and Certifica 2 categories; 3- For the S1 and S2 categories, a term of conformity is issued stating that production followed the MAPA standards and is signed by a Technical Officer (MAPA, 2009).

In high quality seed production technology, processing is an important part of this process, as it aims to remove impurities, undesirable material, seeds from invasive plants, seeds that are under-ripe, deteriorated seeds, poorly formed seeds and seeds that have been attacked by pests and fungi. It is worth remembering that this separation is only possible when there are physical differences between the seed and the undesirable material. Therefore, processing can provide seed lots with better physiological, health and physical quality (DESCHAMPS, 2006). In summary, processing is a set of operations in which the seed passes through the seed processing unit until it is distributed to the producer. All these stages aim to improve the physical characteristics of the seeds, where the function is to seek maximum quality for production conditions in the field (PESKE and BAUDET, 2003).

2.3 SEED PATHOLOGY

Seeds, as the main input, are given greater importance by any agricultural sector due to the various microorganisms associated with them, which can be a highly undesirable factor in the initial establishment of a crop (GOULART, 1997). The guarantee of better establishment in a given area depends on the use of highly vigorous, disease-free seeds (YORINORI, 1988).

According to PERETTI (1994), quality seeds are those that have high viability, i.e. are capable of producing normal plants under unfavorable environmental conditions, which can occur in the field. For a seed lot to be considered quality, it must belong to the desired species and cultivar; be pure, i.e. contain no other seeds or inert materials; have no dormancy and, if it does, this must be naturally reversible; have a high level of germination and excellent health; be easy to preserve, i.e. have a low water content, and be well adapted to the edaphic and climatic conditions of the region for which it is intended.

The sanitary condition of soybean seeds is of the utmost importance if we consider that they are carriers of pathogens, which lodge in them and are then taken to the field, causing the first foci of infection, thus reducing the germination potential and vigor of the seeds (GOULART, 1997).

Soybeans are attacked by a wide range of fungal diseases, some bacteria, as well as nematodes and viruses, the fungi group being the most common. Most of these organisms use seeds as the main vehicle for dissemination and introduction into new areas of cultivation, which, under favorable environmental conditions, can cause damage to the crop (HENNING, 1987 and 1996).

In order to verify whether or not a pathogen is present in seeds or plant parts and is the same causal agent of a disease in the host species in question, pathogenicity tests are necessary. These are relatively quick and easy methods and can be carried out without the use of sophisticated equipment, leading to faster diagnosis and control measures (NECHET; ABREU, 2002).

The pathogens that cause infection in seeds return to the aerial or root organs. This is a very important process because it allows the pathogen to continue its life cycle by providing an essential nutritional source for its growth and sporulation. The fungus can remain viable until the next crop is sown, as long as infected seeds are stored. During the seed germination process, the mycelium that was in the pericarp, endosperm or embryo starts growing again. They are very responsive to environmental stimuli, in this case water. As the seed contains 12 to 13 % moisture during storage, this favors an adverse environment for the vegetative growth of infecting pathogens. At this moisture content, both the pathogen and the seed are dormant. In the soil, the seed is hydrated when it comes into contact with soil water. At this stage, the mycelium takes on a vital activity and begins to grow towards the surface of the seed (DHINGRA, 2005).

According to Dhingra (2005), pathogens spread in the following ways: 1- internal pathogen, which is located within some tissue of the seed, including the pericarp; 2- external pathogen, which is passively adhered to the seed coat; and 3- joint contamination, in which case the pathogen is not associated with the seed, but accompanies it along with the seed mass in the form of propagules, such as sclerotia, mycelium or resistance spores, located within the remains of the root, leaf, stem, fruit and soil.

According to Oliveira et al. (2005), seed transmission is closely linked to survival and dissemination. The longevity of the primary inoculum is extremely important for bacterial pathogens. Survival depends on their ability to escape from adverse conditions, including environmental ones. Phytobacteria have a survival mechanism in which they are associated with the seeds of their hosts. The seed provides a good substrate for their survival and development. The transportation of this pathogen by the seed is a very important condition for its survival and the transfer of inoculum over time.

According to Machado & Pozza (2005) some of the risks involved in using infected seeds are a reduction in the vigor and germination power of the seeds; accumulation of inoculum in the growing areas; greater need for applications of phytosanitary products; reduced productivity; selection of more aggressive/ virulent populations; rendering the area unusable for growing certain plant species.

It is important to emphasize that the losses associated with pathogens in seeds are not limited to reductions in plant stand, but rather a series of factors that can cause immeasurable damage to the entire agricultural system.

Currently, seed multiplication companies offer their customers seed treatment in advance, with the aim of helping them by guaranteeing greater practicality and speed at the time of sowing. Chemical treatment is a technique that can be used for any quantity of seed, as various pieces of equipment have been developed and improved in recent years (BAIL, 2013). Existing seed treatment machines have the advantage of higher yields (60 to 70 bags per hour), guaranteeing greater coverage, product adherence and reducing the risk of operator poisoning (HENNING et al, 2004).

2.4 PHYSIOLOGICAL QUALITY

According to POPINIGIS (1977) and PESKE & BARROS (1998), the physiological quality of seed is the sum of its attributes that indicate its ability to perform vital functions, such as germination, vigor and longevity. The use of seeds with high physiological quality directly influences the development of the crop, providing greater population uniformity, absence of seed-borne diseases, high plant vigor and high productivity.

In the seed formation process, the point of maximum physiological quality is considered to be the point at which the seed shows maximum germination and vigor. As, in general, at this point the seed's humidity is considered too high for harvesting to take place, the harvest is postponed and the seeds remain stored in the field, exposed to environmental conditions (CARVALHO & NAKAGAWA, 2000).

One of the most important stages in soybean production is obtaining high quality and suitable seeds, so that these can be used economically by producers when sowing their crops, guaranteeing their development (ROCHA et al., 1996).

Seed vigor is one of the main physiological quality attributes to be considered when planting crops. However, due to its complexity, vigor cannot always be fully assessed by just one test, which is why it is recommended to use several tests to get a more accurate idea of the physiological quality of a seed lot (SCHEEREN et al., 2010). Seeds with low vigor can cause reductions in speed and total emergence, initial size, dry matter production, leaf area and plant growth rates (KOLCHINSKI et al., 2005). Batches with lower vigor, due to greater variation between the seeds, show greater unevenness and lower speed of emergence (SCHUCH et al., 1999).

It is common for a batch to show variations in physiological quality between seeds. Batches with lower vigor, due to the greater variation between seeds, show greater unevenness and lower speed of emergence. Schuch et al. (1999) found that a reduction in the level of seed vigor increased the average time required for root protrusion, as well as reducing the average number of roots emitted per day. The greater speed of emergence and the production of larger seedlings may give plants from vigorous seeds an initial advantage in the use of water, light and nutrients.

The physiological quality of the seeds can affect their ability to regenerate the plant. Effects of seed physiological quality on the speed and uniformity of emergence, total emergence and plant establishment have been documented. MACHADO (2002) found that the progressive reduction in the physiological quality of seeds led to reduced and uneven emergence in the field.

The physiological quality of soybean seeds is largely due to the influence of the genotype. In recent years, breeding programs have been concerned with developing materials with greater resistance to diseases and pests, oil and protein content (Costa, N. et al. 2001).

Several factors can influence the physiological quality of soybean seed, which can occur during the production phase in the field, harvesting, drying, storage, processing and sowing. These factors include temperature extremes during ripening, fluctuations in ambient humidity conditions, including droughts, deficiencies in plant nutrition, the occurrence of insects and diseases, as well as the adoption of inadequate harvesting, drying and storage techniques (KRZYZANOWSKI et al., 2008).

The variation in humidity, associated with the delay in harvesting, can alter the gain and loss of water by the seeds due to their hygroscopicity, causing various damages such as wrinkling of the tegument, thus accelerating the deterioration process, due to the fact that it makes it easier for pathogens to enter and the embryo is more exposed to the environment (SANTOS et al., 1996).

Some authors have used water immersion tests on seeds and vegetables to assess seed quality. The impermeability of the seed coat is considered to be a mechanism of tolerance to seed deterioration. This characteristic is shown mainly by hard seeds, which make it difficult for water to penetrate the seed coat (Custódio et al., 2002).

To measure whether soybean seeds have good physiological quality, germination and vigor parameters are used, where the vigor of the seeds, added to some attribute, confers their germination potential, thus resulting in normal seedlings under various environmental conditions (HOFS ET AL.,2004).

The physiological quality of soybean seeds is largely influenced by the genotype. In recent years, breeding programs have sought to develop materials with characteristics such as resistance to diseases and pests, oil and protein content and, more recently, lignin content in the seed coat (COSTA et al., 2001). However, factors such as the delay in harvesting soybeans after physiological maturity can cause reductions in seed germination and vigor, depending on genetic factors and the conditions of the natural environment to which they are exposed (MINUZZI et al., 2010).

The germination test is used to measure the quality of a batch of seeds. This was developed to serve the seed trade and is a standardized test, thus allowing it to be reproduced. It is unlikely to be able to predict seed performance in the field, as it is carried out under optimum laboratory conditions (MARCOS FILHO, 2005).

As recommended, the use of foliar fungicides can lead to better yields and high seed quality (PESKE and BARROS, 2006). According to reports by Klingelfuss and Yorinori (2001), there is still a lack of information on the residual period of fungicides in relation to end-of-cycle diseases.

2.5 damages and economic losses

Within Brazilian agribusiness and that of several countries, soya occupies a prominent place, but with the rapid expansion of pathogens, significant economic and technical damage has occurred, which has greatly compromised the profitability of producers and consequently the economy of these countries (YORINORI; LAZZAROTTO, 2004).

The economic importance of each disease can vary from year to year and from region to region, depending on the weather conditions of each harvest (EMBRAPA, 2000). In soybeans, annual losses due to diseases are estimated at around 15% to 20%. However, some diseases cause losses of almost 100% (EMBRAPA, 2007).

When seed quality is compromised, it becomes more difficult to obtain a uniform plant stand, which ends up directly influencing crop production. In some situations where the plant population is below that recommended for the cultivar, there will be a need to carry out reseeding, a practice that includes an increase in production costs and the inherent risks that this practice can bring, such as sowing outside the recommended period, a possible change of cultivar and problems with fertilization, factors that cause losses for the producer and reduce productivity (KRZYZANOWSKI, et al., 2008).

With the opening up of new agricultural frontiers, mainly due to the great expansion of soybeans in these areas, without the correct adoption of technical criteria, there has been a rapid spread of diseases in Brazil's soybean crop. Due to this spread of diseases in the crop, some parameters have been significantly affected: grain quality and productivity, consequently increasing soybean production costs, mainly due to the increase in fungicide applications. Some diseases worth mentioning are Asian rust (*Phakopsora pachyrhizi*), anthracnose (*Colletotrichum truncatum* var. trucata), oidium (*Microsphaera diffusa*), mildew (*Peronospora manshurica*), end-of-cycle diseases (*Septoria glycines* and *Cercospora kikuchii*), sclerotium wilt (*Sclerotium rolfsii*), target spot (*Corynespora cassiicola*) and mela or soybean late blight (*Rhizoctonia solani)* (EMBRAPA SOJA, 2002).

According to Peske et al. (2006), there have been reports that seed deterioration in the field can interact with some fungi, including *Phomopsis spp. and Colletotricum truncatum,* which cause a reduction in seed vigor and germination, thus damaging the establishment of plant stands.

2.6 CHEMICAL GROUPS

The main chemical groups used in soybean cultivation to control oidium, rust and end-of-cycle diseases are triazoles and strobilurins. For the control of bean rust (*Uromyces appendiculatus*), fungicides from the carboxanilide group are the most effective.

Fungicides from the triazole group have a systemic action, inhibiting the action of ergosterol synthesis, which inhibits the germination of spores, the formation of the germ tube and the appressorium. Even if the fungus penetrates the treated host tissues, the fungicide acts by inhibiting mycelial growth inside the tissues (Forcelini, 1994). The recently developed triazoles act on the formation of ergosterol, which is an important fungal lipid in the formation of cell membranes. In the absence of this layer, mycelial growth does not occur, as the fungal cell collapses (Juliatti et al., 2004).

Fungicides from the strobilurin group act by inhibiting mitochondrial respiration, interfering in the formation of ATP, blocking the transfer of electrons by the bcl cytochrome complex, through the inhibition of ubihydroquinone-cytochrome c oxidoreductase in the mitochondria (KIMATI, 1995). Strobilurins are highly effective against spore germination, as well as inhibiting the mycelial growth of fungi. These fungicides have both eradicating and protective properties.

The fungicide Pyraclostrobin belongs to the strobilurin group, which includes numerous molecules such as Kresoxim-Methyl, Azoxystrobin and Trifloxystrobin. The mode of action of these molecules in fungi is to inhibit electron transfer at the Qo site in complex III of the mitochondria (FRAC, 2013). These molecules were first used in the 1990s and have since been widely applied against a wide variety of fungal pathogens.

The carboxamide group had fluxapiroxad as its first systemic active ingredient to be discovered. The first molecules in this group acted on diseases caused by basidiomycetes and their narrow spectrum of fungitoxicity included rusts, scab, cankers and *Rhizoctonia solani* (AMORIM; REZENDE; BERGAMIN FILHO, 2011).

3. MATERIAL AND METHODS

This chapter includes the materials and methods used to carry out the experiment, the study area, the time and manner in which the treatments were carried out, as well as the evaluations carried out during and after the experiment.

3.1 CHARACTERIZATION OF THE STUDY AREA

This study was carried out in two stages. The field stage was conducted in the experimental area of the Department of Plant Health Defense at UFSM, where the fungicide applications were carried out, in the Central Depression region of the state of Rio Grande do Sul, at an altitude of 116m, latitude 29°42'24" S and longitude 53°48'42" W. According to the KOEPPEN classification (Moreno, 1961), the region's climate is Cfa. The soil is classified in the Brazilian Soil Classification System as Argissolo Vermelho Distròfico Arnico (EMBRAPA, 2006). The laboratory evaluations were carried out at the Plant Health Clinic of the Plant Health Defense Department of the Federal University of Santa Maria.

3.2 TREATMENTS AND EXPERIMENTAL DESIGN

The experimental design was a randomized block design with eight treatments (Table 1) and four replications. Applications were made using a CO_2-pressurized backpack/hand sprayer with the respective doses recommended for the crop. The experiment was conducted during the 2014/2015 harvest. The cultivar used was Apolo and each plot represented $3m^2$ of area for evaluation. Four applications were made at V5, R1, R1 + 14 days and R1 + 28 days.

Table 1- Active ingredient, trade name, chemical group and dose of the commercial product (L p.c. ha^{-1}) of the fungicides used in the experiment UFSM, Santa Maria, 2015.

Active ingredient	Chemical group	Dose p.c L/ p.c.ha^{-1}
T1 No treatment		
T2 Pyraclostrobin	Anilide and Strobilurin	..0 .. . .,30
T3 Azoxystrobin	strobilurin	0,35
T4 Picoxystrobin	strobilurin + triazole	0,75
T5 Epoxiconazole + Pyraclostrobin	strobilurin + triazole	0,50
T6 Metconazole + Pyraclostrobin	strobilurin + triazole	0,50
T7 Tebuconazole + Trifloxystrobin	strobilurin + triazole	0,50
T8 Azoxystrobin + Cyproconazole	strobilurin + triazole	0,30
T9 Picoxystrobin + Cyproconazole	strobilurin + triazole	0,30

3.3 HEALTH AND PHYSIOLOGICAL QUALITY

3.3.1   GERMINATION

Fifty seeds were used for each replication, sown on sheets of germitest paper moistened to 2.5 times the mass of the paper, and then conditioned in a Biochemical Oxygen Demand (BOD) incubator, where they were kept at a constant temperature of 25°C, with a 12-hour photoperiod. Evaluations were carried out at five and eight days after the start of the test, according to the RAS (Brasil, 2009), with the results expressed as a percentage of normal seedlings.

Figura 1-   Start of germination test.

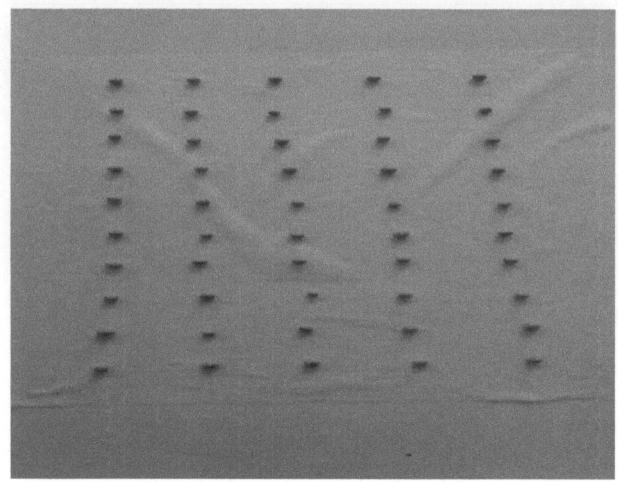

Source: (Boscaini,2015).

Figura 2- Seedlings after eight days.

Source: (Boscaini, 2015).

3.3.2 SEEDLING LENGTH

Fifty seeds per replication were used to evaluate the length of the seedlings. The seeds were allowed to germinate using sheets of moistened paper as a substrate at a temperature of 25°C. Evaluations were carried out eight days after sowing, measuring the length with a ruler graduated in millimeters (aerial part and root) of 20 randomly selected seedlings. The results were expressed as the average length per seedling in centimeters.

Figura 3- Seedling aerial part and root.

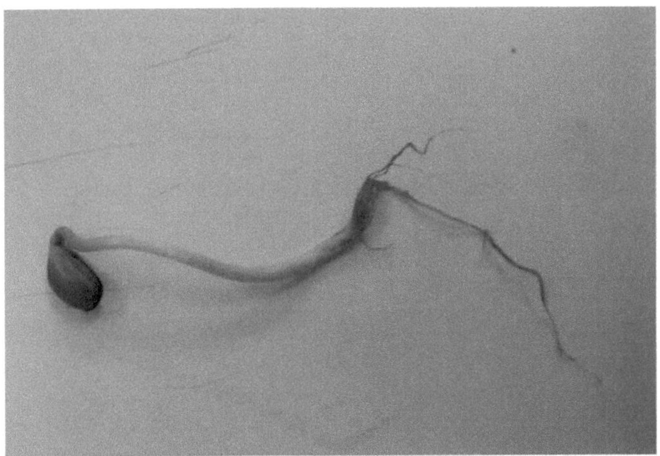

Source: (Boscaini, 2015).

3.3.3 PATHOLOGY

It was carried out using the filter paper method, using 100 seeds per treatment, which were placed on layers of moistened filter paper in *gerboxes* (25 seeds per box). The samples were subjected to a temperature of 25C° with a 12-hour photoperiod for 7 - 8 days. The results were presented as a percentage of infected seeds, where each seed was examined individually with the aid of a magnifying glass and microscope, assessing the occurrence or not of fungal growth.

Figura 4- - Infected seed health test.

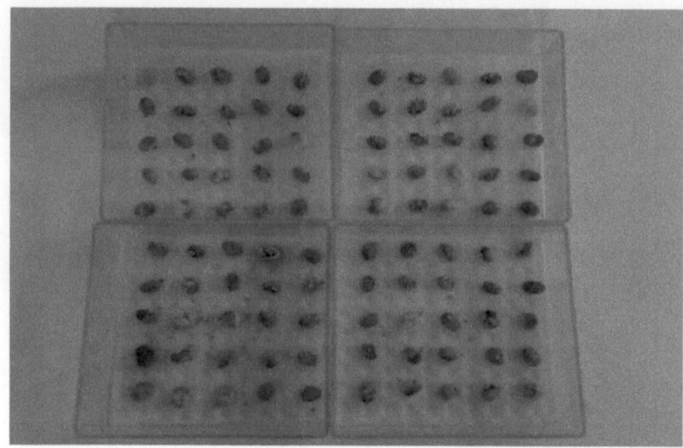

Source: (Boscaini,2015).

Figura 5- - Seeds infected with *Aspergillus sp.,Penicillium sp.,Fusarium sp.*

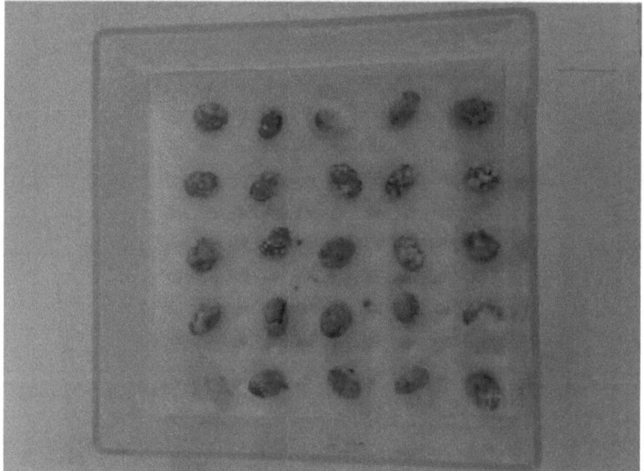

Source: (Boscaini, 2015).

3.3.4 MASS OF 1000 GRAMS

A sample of 1000 grains each was randomly counted from the total number of grains produced by each treatment. The samples were weighed on a 0.001 g precision scale and the average values were expressed in grams and corrected to 13% moisture content. The grain mass values for each treatment were transformed into kg ha^{-1} , and corrected to 13% moisture content.

3.4 STATISTICAL ANALYSIS

In the statistical analysis of the data from the experiment, the variables that were significant using the F test (ANOVA) were compared using the Scott-Knott test, with a 5% probability of error. The program used for data analysis was Sisvar® software (FERREIRA, 2008). For those variables where the data was transformed, the results were represented using the original values.

4. RESULTS AND DISCUSSION

This chapter deals with the results obtained in this experiment, through data analysis, which were discussed and compared with other studies already carried out in the literature.

The results of the evaluation of the physiological quality of soybean seeds are shown in Tables 2, 3 and 4.

The mean comparison test (Table 2) for the seed vigor variable showed that there was a statistical difference between the treatments compared to the control. The pyraclostrobin treatment had the highest germination percentage compared to the other treatments.

However, when comparing the effectiveness of the treatments with each other, the metconazole + pyraclostrobin treatment had the lowest germination percentage and was significantly lower than the other chemical treatments as well as the control, differing statistically at a 5% probability of error.

All the treatments had low vigor, with an overall average of 29.72%, which is below the level recommended by seed companies, where for a batch of seeds to be considered high quality, it must have at least 80% vigor

According to Morais et al. (2008), a probable cause for the low germination performance of seeds can be associated with the significant presence of a large number of storage pathogens. Goulart (2000) considers that treatment with fungicides has no effect on factors such as mechanical damage, deterioration due to humidity, bedbug attacks and inadequate storage, which cause a reduction in the physiological quality of seeds.

Table 2- Average vigor data (V) of soybean seeds from plants subjected to different fungicide applications. UFSM, Santa Maria, 2015.

Treatments	Vigor
Untreated witness	36,50 B

Pyraclostrobin	43,75 A
Azoxystrobin	27,00 C
Picoxystrobin	26,75 C
Epoxiconazole + Pyraclostrobin	29,00 C
Metconazole + Pyraclostrobin	5,75 D
Tebuconazole + Trifloxystrobin	26,75 C
Azoxystrobin + Cyproconazole	36,00 B
Picoxystrobin + Cyproconazole	36,00 B
Overall average	**29,72**

*Means followed by the same letter in the column and within each variable do not differ by the Scott-Knott test, p>0.05.

For the variables normal seedlings, abnormal seedlings and dead seeds, the average data is shown in Table 3. The data obtained in the study for the normal seedling variable showed that there was a significant difference between the treatments compared to the control. The pyraclostrobin, azoxystrobin + cyproconazole and picoxystrobin + cyproconazole treatments, together with the control, obtained the best results in terms of percentage of normal seedlings compared to the other treatments used. The treatment with metconazole + pyraclostrobin had the lowest percentage of normal seedlings, showing that this fungicide, when applied in the field, affects the physiological quality of the seed.

As for the abnormal seedling variable, there was a static difference between the treatments and the control. The treatment with azoxystrobin had the lowest percentage of abnormal seedlings. This shows that the treatment was efficient in controlling diseases in the field, providing seeds with high physiological quality.

With regard to dead seeds, the results show that there was a significant difference between the treatments (Table 3). The treatment with metconazole

+ pyraclostrobin was the one with the highest percentage of dead seeds, differing significantly from the other treatments with fungicide application as well as from the control without application. The treatments with pyraclostrobin and picoxystrobin + cyproconazole had the lowest percentage of dead seeds.

Table 3- Average data for normal seedlings (NP), abnormal seedlings (AP) and dead seeds (SM), soybeans, from plants subjected to applications of different fungicides. UFSM, Santa Maria, 2015.

Treatments	Normal seedlings	Abnormal seedlings	Dead seeds
Untreated witness	41,25 A	2,25 A	6,50 A
Pyraclostrobin	44,25 A	1,25 A	4,50 A
Azoxystrobin	27,75 B	0,00 A	22,50 C
Picoxystrobin	19,00 C	6,50 B	24,50 C
Epoxiconazole + Pyraclostrobin	33,00 B	3,50 B	13,50 B
Metconazole + Pyraclostrobin	0,00 D	5,50 B	44,50 D
Tebuconazole + Trifloxystrobin	29,00 B	0,75 A	20,25 C
Azoxystrobin + Cyproconazole	37,50 A	1,50 A	11,00 B
Picoxystrobin + Cyproconazole	40,25 A	5,25 B	4,50 A
Overall average	**30,22**	**2,94**	**16,86**

*Means followed by the same letter in the column and within each variable do not differ by the Scott-Knott test, p>0.05.

The average data observed for the aerial part and root show that the applications with picoxystrobin + Cyproconazole and the untreated control showed the best performance for the variables that did not differ from each other, where these treatments did not influence the growth of the seedling. On the other hand, the treatment with metconazole + pyraclostrobin was the one that showed the lowest rate of both aerial and root development, showing that this treatment inhibited the seed's ability to develop.

Table 4- Average data of the aerial part (AP) and root (R) of soybeans from plants subjected to applications of different fungicides. UFSM, Santa Maria, 2015.

Treatments	Aerial	Root
Untreated witness	7,01 A	7,86 A
Pyraclostrobin	3,74 D	2,78 E
Azoxystrobin	3,12 D	2,86 E
Picoxystrobin	4,68 C	3,86 D
Epoxiconazole + Pyraclostrobin	4,71 C	3,67 D
Metconazole + Pyraclostrobin	0,00 E	0,00 F
Tebuconazole + Trifloxystrobin	5,41 B	3,91 D
Azoxystrobin + Cyproconazole	5,80 B	4,28 C
Picoxystrobin + Cyproconazole	7,12 A	7,32 B
Overall average	**4,62**	**4,06**

*Means followed by the same letter in the column and within each variable do not differ by the Scott-Knott test, $p > 0.05$.

According to the data obtained (Table 5), the pathogens associated with soybean seeds were *Aspergillus sp.*, *Pénicillium sp. and Fusarium sp. The* effect of all the treatments used was observed for the fungus *Aspergillus sp.* The chemical treatments metconazole + pyraclostrobin and tebuconazole + trifloxystrobin were the most effective, as they showed a reduced incidence of the fungus compared to the control, showing that chemical control in the field was effective in eliminating the pathogen. However, the application of azoxystrobin + cyproconazole proved to be the worst treatment, differing statistically from all the other treatments with an average incidence of 45.00% of the pathogen.

For the fungus *Penicillium sp.* There was no effect of the chemical treatment for six of the nine treatments, showing how effective they were in eliminating the pathogen, compared to the control, which had the highest infection rate, due to the fact that no fungicides were applied in this treatment.

As for the fungus *Fusarium sp.*, the evaluations show a lower average infestation of the pathogen for the treatment with metconazole + pyraclostrobin, which was therefore the most effective over the other treatments. However, the control treatment had the highest average infection rate over the other applications, which shows the high contamination of the seeds.

According to Machado (2000), the greater or lesser efficiency of chemical treatment depends greatly on the type of seed, the physical and physiological condition of the seed, the variability of the pathogen, the level of contamination and the active ingredient and dosage of the product.

Table 5- Percentage of infection (health) for the fungi *Aspergillus sp.*, *Penicillium sp.* and *Fusarium sp.* in soybean seeds from plants subjected to applications of different fungicides. UFSM, Santa Maria, 2015.

Treatments	*Aspergillus sp.*	*Pénicillium sp.*	*Fusarium sp.*
Witness without treatment	26,75 C	23,7 C	78,00 E
Pyraclostrobin	20,00 B	16,00 B	18,85 B
Azoxystrobin	16,50 B	5,75 A	23,30 C
Picoxystrobin	29,00 C	5,75 A	31,00 D
Epoxiconazole + Pyraclostrobin	20,15 B	2,90 A	21,85 B
Metconazole + Pyraclostrobin	1,75 A	1,60 A	4,00 A
Tebuconazole + Trifloxystrobin	5,15 A	4,00 A	18,30 B
Azoxystrobin + Cyproconazole	45,00 D	14,90 B	26,40 C
Picoxystrobin + Cyproconazole	17,00 B	5,37 A	25,50 C
Overall Average	**20,14**	**8,88**	**27,46**

*Means followed by the same letter in the column and within each variable do not differ by the Scott-Knott test, p>0.05.

In the determinations carried out in the field, the variables grain yield (kg ha^{-1}) and yield (sc ha^{-1}), and thousand-grain mass (g), there was a significant effect of the chemical treatments only for the thousand-grain mass variable.

For the grain yield and bag/ha yield variables, the data obtained in the mean comparison test showed no significant effect between the different applications. Rezende et al. (2003), evaluated that even though there was a significant effect of the chemical treatment of soybean seeds on plant emergence in the field, the same was not observed for the final yield of the crop.

Table 6 shows that there was a significant difference between the treatments with pyraclostrobin and tebuconazole + trifloxystrobin. This shows that the various treatments used in the field did not show any significant difference, and these did not affect the final yield of the crop.

Table 6- Average final grain yield (kg. ha^{-1}), yield (Sc.ha^{-1}) and thousand grain mass (g) of soybean seeds from plants subjected to applications of different fungicides. UFSM, Santa Maria, 2015.

Treatments	Income		Mass of one
	(kg.ha)$^{-1}$	(Sc.ha^{-1})	thousand grams (g)
Witness without treatment	559,61 A	9,33 A	157,27 B
Pyraclostrobin	620,37 A	10,34 A	170,50 A
Azoxystrobin	636,43 A	10,61 A	159,23 B
Picoxystrobin	631,36 A	10,52 A	162,26 B
Epoxiconazole + Pyraclostrobin	664,80 A	11,08 A	164,34 B
Metconazole + Pyraclostrobin	728,53 A	12,14 A	156,68 B
Tebuconazole +	637,87 A	10,63 A	177,96 A

Trifloxystrobin			
Azoxystrobin +			
Cyproconazole	814,03 A	13,57 A	157,22 B
Picoxystrobin +			
Cyproconazole	559,61 A	9,33 A	156,54 B
Overall average	**650,29**	**10,84**	**162,44**

*Means followed by the same letter in the column and within each variable do not differ by the Scott-Knott test, p>0.05.

5. CONCLUSIONS

Based on the results, the following conclusions were reached:

The chemical treatment with picoxystrobin + cyproconazole was the most effective for the physiological evaluations (normal seedlings, abnormal seedlings, dead seeds and vigor) of soybean seeds;

The chemical treatment metconazole + pyraclostrobin gave greater control of pathogens associated with the seeds;

The fungicides used did not differ in terms of soybean yield.

6. BIBLIOGRAPHICAL REFERENCES

ABRASEM. Agriculture without frontiers - Brazil generating technology and food. Pelotas: Editora Becker & Peske Ltda., 2011.

AMORIM, L.; REZENDE, J.A.M.; BERGAMIN FILHO, A. **Manual de Fitopatologia**. 4. ed Piracicaba: Agronômica Ceres, 2011. 704 p.

ALMEIDA, A.M.R.; FERREIRA, L.P.; YORINORI, J.T.; SILVA, J.F.V.; HENNING, A.A. GODOY, C.V.; COSTAMILAN, L.M.; MEYER, M.C. Soybean diseases. In: KIMATI, H.; AMORIM, L.; REZENDE, J.A.M.; BERGAMIN FILHO, A.; CAMARGO, L.E.A. **Manual de Fitopatologia.** Diseases of cultivated plants. 3. ed. v.2. Sâo Paulo: Agronòmica Ceres, 2005. p.569-588.

BAIL, J.L. **Relationships between soybean seed treatment, physiological and health parameters and seed preservation**. 2013. 41f.

Dissertation (Master's Degree in Agronomy). State University of Ponta Grossa. 2013.

BALARDIN, R. S.; SILVA, F. D. L.; CORTE, G. D.; FAVERA, D. D.; TORMEN, N. R.. Seed treatment with fungicides and insecticides as reducers of the effects of water stress on soybean plants. **Ciência Rural**, Santa Maria, v. 41, n. 7, p. 1120-1126, 2011.

BARBOSA, C.Z.R.; SMIDERLE, O.J.; ALVES, J.M.A.; VILARINHO, A.A.; SEDIYAMA, T. Quality of BRS Tracajà soybean seeds, harvested in

Roraima as a function of storage size. Revista Ciência Agronômica, v.41, n.1, p.73-80, 2010.

BARROS, R. G.; BARRIGOSSI, J. A. F.; COSTA, J. L. S. Effect of storage on the compatibility of fungicides and insecticides, associated or not with a polymer in the treatment of bean seeds. **Bragantia**, Campinas, v. 64, n. 3, p. 459-465, 2005.

BAUDET, L.; PESKE, F. Increasing seed performance. Seed News, Pelotas, v. 9, n. 5, p. 22-24, 2007.

BRAZIL. Law No. 10.711, of August 5, 2003. Provides for the National Seed and Seedling System and other measures. Official Gazette of the Union, August 5, 2003.

BRAZIL. Ministry of Agriculture and Agrarian Reform. **Regras para anàlisedesementes**,Brasilia,2009.395p.Available at:http://www.agricultura. gov.br/arq_editor/file/2946_regras_analise__sementes.pdf. Accessed on: December 4, 2015.

BRAZIL. Ministry of Agriculture, Livestock and Supply. **Rules for seed analysis**. Ministry of Agriculture, Livestock and Supply. Secretariat for Agricultural Defense. Brasilia, DF: Mapa/ACS. 2009, 395 p.

BUZZERIO, N. F. Seed quality tools in professional seed treatment. **ABRATES Newsletter**, v. 20, n. 3, p. 56, 2010.

CARVALHO, N.M. & NAKAGAWA, J. **Seeds: science, technology and production.** Jaboticabal: Funep, 4. ed., 2000, 588p.

CONAB. National Supply Company. **Crops**. Captured on December 15, 2015. Online. http://www.conab.gov.br.

COSTA, N.P. da; MESQUITA, C. de M.; MAURINA, A.C,; FRANÇA NETO, J. de B.; PEREIRA, J.E.; BORDINGNON, J.R,; KRZYZANOWSKI, F.C.; HENNING, A.A. Efeito da colheita mecânica da soja nas Caracteristicas fisicas, fisiológicas e quimicas das sementes em três estados do Brasil. **Revista Brasileira de Sementes**, Brasilia, v.23, n.1, p.140-145, jan./feb. 2001.

COSTA, N.P.; MESQUITA, C.M.; MAURINA, A.C.; FRANÇA NETO, J.B.; PEREIRA, J.E.; BORDINGNON, J.R.; KRZYZANOWSKI, F.C; HENNING, A.A. Effect of mechanical soybean harvesting on the physical, physiological and chemical characteristics of seeds produced in three Brazilian states. Revista Brasileira de Sementes. Londrina, v.23, n.1, p.140-145, 2001.

CUSTÓDIO, C.C.; MACHADO NETO, N.B.; MASSAKI ITO, H.; VIVAN, M.R. Effect of water submergence of bean seeds on germination and vigor. **Revista Brasileira de Sementes**, Brasilia, v.24, n.2, p. 49-54, mar./abr.2002.

DESCHAMPS, L.H. Qualidade da semente de soja e de seu repasse beneficiados em mesa de gravidade - Pelotas, 2006. - 46f. : il. Dissertation (Master's Degree). Seed Science and Technology. Eliseu Maciel School of Agronomy. Federal University of Pelotas, Pelotas, 2006.

DHINGRA, O. D. Theory of the transmission of fungal pathogens by seeds. In: ZAMBOLIM, L. Seeds: phytosanitary quality. Vicosa: UFV, 2005.

DIAS, W.P.; ASMUS, G.L.; SILVA, J.F.V.; GARCIA, A.; CARNEIRO, G.E.S. Nematodes. In: Almeida, A.M.R.; Seixas, C.D.S.(Ed.) Soybeans: root and stem diseases and interrelationships with soil and crop management. Embrapa Soja: Londrina, 2010. P. 173-2

EMBRAPA (2003) Soybean Production Technologies Central Region of Brazil 2003. Londrina: Embrapa Soja.

EMBRAPA SOJA. (Brazil). Soybean production technologies - central region of Brazil - 2007. - Londrina: EMBRAPA - CNPSo/Embrapa

Cerrados/Embrapa Agropecuâria Oeste/Embrapa Soja, 2006. n.11. p.169218.

EMBRAPA. Soybean cultivation in Brazil. Londrina: EMBRAPA Soja, 2000. 179p.

EMBRAPA. Soybean production technologies - central region of Brazil - 2007. Londrina: Embrapa Soja: Embrapa Cerrados: Embrapa Agropecuâria Oeste, 2007, 225p.

EMBRAPA-Brazilian Agricultural Research Corporation. **Brazilian soil classification system.** 2. ed. Rio de Janeiro: Embrapa Solos, 2006. 306 p.

FAO. Seed production. Available at: Accessed on: 18 Nov 2015.

FERREIRA, D. F. SISVAR: a program for statistical analysis and statistics teaching. **Revista Symposium**, Recife, v. 6, p. 36-41, 2008.

FORCELINI, C.A. Fungicides that inhibit sterol synthesis. Triazoles. **Annual Journal of Plant Pathology**, v.2, n.1, p.335-351, 1994.

FERREIRA, R.L. Stages of processing in the physical and physiological quality of corn seeds. Unesp, Ilha Solteira, 2010. 49f. (Master's dissertation).

FRAC. **FRAC Code List 2013**: Fungicides sorted by mode of action. 2013. 10p.

GIANASI, L. et al. Efficiency of captan fungicide associated with other fungicides in the chemical treatment of soybean seeds. Summa Phytopathologica, v. 26, n. 02, p. 241-245, 2000.

GOULART, A. C. P. Efficiency of different fungicides in controlling pathogens in soybean seeds and their effects on the emergence and grain yield of the crop. **Informativo Abrates**, Londrina, v. 10, n. 1/2/3, p.17-24, 2000.

GOULART, A. C. P.; FIALHO, W. F. B.; FUJINO, M. T. **Viabilidade Técnica do Tratamento de Sementes de Soja com Fungicidas antes do Armazenamento.** Informativo ABRATES, v.9, n. 1 / 2, p.110, 1999.

GOULART, A. C. P. Fungi in Soybean Seeds: Detection and Importance.Accessed30.01.2016.Availableat:https://www.infoteca.cnpti a.embrapa.br/infoteca/bitstream/doc/240627/1/doc1197.pdf

HENNING, A.A. **Pathology and seed treatment.** General notions. Londrina: EMBRAPA-SOJA, 2004. 51p. Documents n.235.

HIN, C.J.A. Market prospects for sustainable soybeans in the Netherlands. **CLM Onderzoek en Advies BV** (Research Center for Agriculture and the Environment) Utrecht, Netherlands. 2002.

HOFS, A.; SCHUCH, L.O.B.; PESKE, S.T. BARROS, A.C.S.A. Emergence and growth of rice seedlings in response to seed physiological quality. **Revista Brasileira de Sementes**, Pelotas, v.26 n.1, p.92-97, 2004.

JULIATTI, F.C.; APPELT, C.C. N.S.; BRITO, C.H.; GOMES, L.S.; BRANDÂO, A.M.; HAMAWAKI, O.T.; MELO, B. Control of phaeospheridium, common rust and cercosporiosis by the use of genetic resistance, fungicides and times of application in corn cultivation. **Bioscience Journal**, v.20, n.3, p.45-54, 2004.

JULIATTI, F. C. Advances in chemical seed treatment. **Informativo ABRATES,v.20,n.3,**p.54-55,2010.

Kiihl Ra S (2006) How to get the "locomotive" back on track? Visâo Agricola, USP-ESALQ, year 3, p. 4-7.

KIMATI, H. Chemical control. In: BERGAMIN FILHO, A.; KIMATI, H.;

AMORIM, L. (Ed.) **Manual de Fitopatologia: Principios e Conceitos.** Sâo Paulo, Editora

Agronômica Ceres. v. 1, 3 ed., 1995. p.761-785.

KINGELFUSS, L.H.; YORINORI, J. T. Latent infection of *Colletotrichum truncatum* and *Cercospora kikuchii* in soybean. **Fitopatologia Brasileira**, v.26, n.2, 2001.

KOLCHINSKI, E.M.; SCHUCH, L.O.B.; PESKE, S.T. Seed vigor and intraspecific competition in soybean. Ciência Rural. Santa Maria, v.35, n.6, p.1248- 1256, 2005.

KROHN, N. G.; MALAVASI, M. M. Physiological quality of soybean seeds treated with fungicides during and after storage. Revista Brasileira de Sementes, Pelotas, v. 26, n. 2, p. 91-97, 2004.

KRZYZANOWSKI, F. C.; FRANÇA NETO, J. de B.; HENNING, A. A.; COSTA, N. P.da. **Quality control adding value to soybean seed - seed series.** Londrina: Embrapa Soja, 2008. 12 p. (Embrapa Soja. Circular técnica,54).

LAZAROTO, A.; SANTOS, I.; KONFLANZ, V.; MALAGI, G.; CAMOCHENA, R. C. Diagrammatic scale for assessing the severity of common helminthosporiosis in maize. Ciência Rural, Santa Maria, v. 42, n. 12, p. 2131-2137, 2012.

LUDWIG, P. M.; FILHO, O. A. L.; BAUDET, L.; DUTRA, L. M. C.; AVELAR, S. A. G.; CRIZEL, R. L. Quality of stored soybean seeds after coating with amino acid, polymer, fungicide and insecticide. Revista Brasileira de Sementes, Londrina, v. 33, n. 3, p. 395-406, 2011.

MACHADO, J. C. **Seed treatment for disease control.** Lavras. LAPS/UFLA/FAEPE, 2000, 138 p.

MACHADO, R.F. **Performance of white oats (Avena sativa L.) as a function of seed vigor and plant population. 2002. 46f.**

Dissertation (Master's Degree in Seed Science and Technology) - Postgraduate Course in Seed Science and Technology, Federal University of Pelotas.

MACHADO, J.C; POZZA, E. A. Reasons and procedures for establishing pathogen tolerance in seeds. In: ZAMBOLIM, L. Seeds: phytosanitary quality. Vicosa: UFV, 2005.

MAFIA, R. G.; FERREIRA, M. A.; BINOTI, D. H. B.; LEITE, H. G. Statistical procedure for validating diagrammatic scales in disease quantification. Revista Arvore, Viçosa, v. 35, n. 2, p. 199-204, 2011.

MAPA - Ministry of Agriculture, Livestock and Supply. Accessed on 23.11.2015. Available at: http://www.agricultura.gov.br.

MARCOS FILHO, J. **Fisiologia de sementes de plantas cultivadas.** Piracicaba:Fealq, 2005. v.12. 495p.

MERTZ, L. M.; HENNING, F. A.; ZIMMER, P. D. Bioprotectors and chemical fungicides in soybean seed treatment. **Ciência Rural**, Santa Maria, v. 39, n. 1, p. 13-18, 2009.

MINISTRY OF AGRICULTURE, LIVESTOCK AND SUPPLY. Rules for seed analysis. Brasilia, 2009.

MORAIS, L. A. S; RAMOS, N; BETTIOL, W; CHAVES, F. C. M. **Effects of essential oils on soybean seed germination and health.** Congresso Brasileiro de Olericultura, 48. Horticultura Brasileira 26, 2008.

MORENO, J. A. **Climate of Rio Grande do Sul.** Porto Alegre, Department of Agriculture, 1961. 42 p.

MUNIZZI, A; BRACCINI, A.L.; RANGEL, M.A.S; SCAPIM; C.A; ALBRECHT, L.P. Seed quality of four soybean cultivars, harvested in two locations in the state of Mato Grosso do Sul. Revista Brasileira de Sementes, v.32, n.1, p.176-185, 2010.

NECHET, K.L.; ABREU, M.S. Morphological characterization and pathogenicity tests of isolates of Colletotrichum sp. obtained from coffee trees. Ciência e Agrotecnologia, Lavras, v. 26, n. 6, p. 1135-1142, nov./dez. 2002.

NETO, P. et al. Agribusiness Projections. Accessed on 05.12.2015. Available at:http://www.agricultura.gov.br/arq_editor/PROJECOES_DO_AG RONEGOCIO_2025_WEB.pdf.

OLIVEIRA, J. O; MOURA, A. B; SOUZA,R.M. Transmission and control of phytobacteria in seeds. In: ZAMBOLIM, L. Seeds: phytosanitary quality. Vicosa: UFV, 2005.

PADUA, G.P.; KAZUHIKO, R.Z.; ARANTES, N.E.; FRANÇA, J.B.N. Influence of seed size on the physiological quality and yield of soybean crops. Revista Brasileira de Sementes, v. 32, n. 3 p. 009-016, 2010.

PEREIRA, C. E. et al. Performance of soybean seeds treated with fungicides and filmed during storage. Ciência e Agrotecnologia, v. 31, n. 03, p. 656-665, 2007.

PESKE, S. T.; BARROS, A. C. S. A.. Seed production. In: PESKE, S. T.; LUCCA FILHO, O. A.; BARROS, A. C. S. A. **Seeds: scientific and technological foundations**. 2 ed. Pelotas: UFPel, 2006. p.470p.

PESKE, S.T.; BAUDET, L. Seed processing training for Coopervale UBS managers. Abelardo Luz, July 2003. 45p.

PERETTI, A. **Manual para análisis de semillas**. 1ª ed. Buenos Aires: Hemisfério Sur, 1994, 282p.

PESKE, S.T. and BARROS, A.C.S.A. Produçâo de Sementes, Course in Seed Science and Technology, **ABEAS**, 1998. 76p.

POPINIGIS, F. **Seed physiology**. Brasilia: Agiplan, 1977, 289p.

ROCHA, V.S. et al. The quality of soybean seed. Viçosa: Federal University of Viçosa, 1996 (Bulletin, 188).

SANTOS, V.L.M; SILVA, A.A.C; CARDOSO, A.A; SEDIYAMA, T. Evaluation of the quality of soybean seeds (Glycine Max (L.) Merrill) harvested in two seasons using sub-optimal and supra-optimal temperatures. **Revista Brasileira de Sementes**,v.18, n.1, p. 57-62, 1996.

Sediyama T, Pereira MG, Sediyama CS, Gomes JLL (1985) Soybean culture: Part I. Viçosa: UFV, 96p.

Seixas CDS, Godoy CV, Ferreira LP, Yorinori JT, Henning AA, Almeida AMR (2006) Management of soybean diseases in the south and southeast regions. Fitopatologia Brasileira, Brasilia, v.31, p.60-61.

SILVA, L. H. C. P.; CAMPOS, H. D.; SILVA, J. R. C. Management of soybean white mold. In: SILVA, L. H. C. P.; CAMPOS, H. D.; SILVA, J. R. C. (Ed.). Phytosanitary management of agro-energy crops. Lavras: UFLA, p. 205-214, 2010.

SOUTHGATE, D. **Population growth, increased agricultural production and food price trends**. The Electronic Journal of Sustainable Development, 2009.

SCHUCH, L. O. B. **Seed vigor and physiological aspects of production in black oats** (Avena strigosa Schreb). 1999. 127f. Thesis (Doctorate in Seed Science and Technology) - Eliseu Maciel School of Agronomy, Federal University of Pelotas, Pelotas.

SCHEEREN, B.R.; PESKE, S.T.; SCHUCH, L.O.B.; BARROS, A.C.A. Physiological quality and productivity of soybean seeds. Revista Brasileira de Sementes, v.32, n.3, p.35-41, 2010.

SCHUCH, L.O.B. Seed vigor and aspects of production physiology in black oats (Avena strigosa Schreb.). 1999. 127f. Federal University of Pelotas (Doctoral thesis in Seed Science and Technology).

YORINORI, J.T.; NUNES JUNIOR, J.; LAZZAROTTO, J.J. **Asian soybean rust in** Brazil: evolution, economic importance and control. Londrina: Embrapa Soja, 2004. 36p.

Printed by Books on Demand GmbH, Norderstedt / Germany